Gaining the High Ground
over Evolutionism
Workbook

Gaining the High Ground over Evolutionism
Workbook

Robert J. O'Keefe

iUniverse, Inc.
Bloomington

Gaining the High Ground over Evolutionism
Workbook

Copyright © 2012 by Robert J. O'Keefe

RSV scripture quotations from The Catholic Edition of the Revised Standard Version of the Bible, copyright © 1965, 1966 National Council of the Churches of Christ in the United States of America. Used by permission. All rights reserved.

iUniverse books may be ordered through booksellers or by contacting:

iUniverse
1663 Liberty Drive
Bloomington, IN 47403
www.iuniverse.com
1-800-Authors (1-800-288-4677)

ISBN: 978-1-4759-4965-0 (sc)
ISBN: 978-1-4759-4966-7 (e)

Printed in the United States of America

iUniverse rev. date: 10/17/2012

Contents

Questions

Questions are arranged by chapter and consist of true/false (T / F), multiple choice, fill-in-the-blank, and essay-style questions for which short phrases or sentences are often, though not always, sufficient to answer. Some multiple choice questions may have more than one correct answer. Also, a few questions have been added as discussion questions for which there is no one correct answer; these questions are noted as such in the answer key.

Chapter 1
Questions:

1. T / F

____ Science enables us to understand the natural world.

____ Science has always been the means of understanding the natural world.

____ Aristotle's four categories of causes were known collectively as the "systematic philosophy."

____ Christian theology and natural philosophy were incompatible before the scientific revolution.

____ Francis Bacon advocated an experimental method for verifying theories about natural phenomena.

____ Nicolaus Copernicus first theorized that Earth was the center of the universe.

____ Galileo published *On the Revolutions of the Heavenly Spheres* in 1543.

____ Isaac Newton published his theory of universal gravitation in 1687.

____ Isaac Newton started the scientific revolution with *Mathematical Principles of Natural Philosophy*.

____ Natural philosophers sometimes regarded themselves as "priests" of nature interpreting the "book" of nature.

____ Galileo first understood that planetary motion could be described mathematically.

____ The argument for the existence of God from design in nature is called the theological argument.

____ Galileo ran afoul of church authorities in regard to whether or not God's commandments were arbitrary.

____ Deism became a fashionable theology in the eighteenth century.

_____ Neither the Islamic Empire nor China ultimately accepted the idea that nature is rationally ordered.

_____ The uniformity principle enabled science to extend the scientific method into the past to explain the past.

_____ Deism came about as a compromise between theism and atheism.

_____ Deism is associated with the ideas of supremacy of natural laws and uniformity of natural processes.

_____ A rational approach to understanding nature eventually superseded the need for natural laws.

_____ The experimental method requires empirical evidence in support of scientific theories in order for them to be considered valid.

_____ Generalizations and unifying ideas that form the basis of scientific theories are always products of the scientific method.

2. Which event or trend laid the groundwork for a rational approach to understanding nature?

    a. The Protestant Reformation
    b. Adaptation of the dialectic method of reasoning
    c. Translation of the writings of Aristotle and Plato into Latin
    d. Contact with the scientific achievements of the Islamic Empire
    e. Invention of the printing press
    f. Adoption of the rationality of Greek philosophy in Christian theology

3. Which trends characterized the European Renaissance in the sixteenth century?

    a. Increases in geographical exploration and discovery
    b. Fragmentation of what was perceived to be religious authority
    c. Mass production of books in common languages
    d. Realization of mankind's creative and intellectual capacities
    e. Achievements in art, music, architecture, and literature
    f. Adoption of the rationality of Greek philosophy in Christian theology

4. The deistic philosophy arose most as a consequence of which of the following?

    a. The mechanical philosophy
    b. The dialectic method
    c. Geocentricity
    d. The laws of nature
    e. The scientific method
    f. Natural theology

5. Natural philosophy was synonymous with which of the following?

    a. Early science
    b. The laws of nature
    c. The experimental method
    d. Greek philosophy
    e. Natural theology
    f. Deistic philosophy

6. Natural philosophers defended their practice against charges of atheism by using which of the following?

    a. The teleological argument
    b. The experimental method
    c. The dialectic method
    d. The cosmological argument
    e. Advanced theology
    f. More effective means of persuasion

7. A theological dispute between which pair of natural philosophers illustrates the extent to which theology influenced early science?

    a. Galileo and Bacon
    b. Copernicus and Kepler
    c. Galileo and Newton
    d. Galileo and Leibniz
    e. Newton and Leibniz
    f. Bacon and Leibniz

8. Scientific inquiry was extended into the past by means of which principle?

    a. Theological
    b. Metaphysical
    c. Mechanical
    d. Universal
    e. Uniformity
    f. Philosophical

9. Which four of the following were Aristotle's categories of causes?

a. Mechanical cause
b. Composition cause
c. Immediate cause
d. Intermediate cause
e. Formal cause
f. Informal cause
g. Final cause

10. The success of science shows that …

a. the natural world is rational and can be figured out.
b. supernatural forces are no longer behind natural laws.
c. observations are essential to validate theory.
d. science is not limited in its potential to explain everything.
e. there is no longer a need to explain why there is rational order.
f. the idea of there being a design to the natural world is superfluous.

11. What approach to understanding the natural world was accepted by Christian theology, and eventually led to the scientific revolution in Western Europe?

12. Which three out of the four causes of Aristotle were discarded by natural philosophers in their studies of nature? What was the result called?

13. What two other developments were necessary to transform natural philosophy into science as we know it today?

14. To what were the properties of objects or the natural forces acting on them attributed during the scientific revolution? Was this a hindrance or an aid to the development of science?

15. What were the two main steps in the experimental or scientific method as formulated by Francis Bacon during the scientific revolution? Why was the innovation of the experimental method so important?

16. What trend or new development in theology did the scientific approach to understanding the natural world produce during the Enlightenment?

17. What verse in the Bible sometimes causes people to believe that an Earth-centered universe is an idea taught in the Bible? Why is this not really a teaching?

18. What was the assumption involving the created order that formed the foundation for natural history?

19. What idea did Western European culture accept that was the key factor in the success of science in Western Europe? How does consideration of the Islamic Empire and China help to clarify that as the key factor?

20. The history of science shows that the answer to the question of whether or not the natural world was designed has depended on what?

Chapter 2
Questions:

1. When interpreting any work of literature, we should do which of the following? (T / F)

____ Allow ideas originating external to the subject text to influence us.

____ Adjust the meaning of words as necessary to arrive at the correct answer.

____ Factor in all text relevant to the question(s) being considered.

____ Acknowledge statements of fact as statements of fact with due consideration of their context.

____ Never presume that an understandable meaning will be forthcoming.

2. What is the cause for suspicion regarding the appearance of various interpretive theories that concurred with an age of creation of several million or more years?

3. Why should the location from which God spoke, where the Spirit of God was, be important? How can that location be substantiated?

4. Consider a verse of Old Testament poetry such as Isaiah 62:1. What characteristic is exhibited here that aids comprehension of Psalm 104 or Job 38?

5. Why not just say that the commands of God were spoken during the first six days, but fulfilled over much longer periods?

6. Why should the word *day* be regarded in chapter 1 as a twenty-four-hour day? Under what circumstances might it mean a longer period?

7. What is the origin of the seven-day week?

8. T / F

____   The Gospel of Luke traces Jesus's physical ancestry back to Adam.
____   A continual process of incremental development is indicated in Genesis 1.
____   In 1 Corinthians 15, Paul discusses Adam in a manner that establishes his symbolic existence in contrast to Jesus's physical existence.
____   Job 38 indicates that initially, the sea waters were covered by clouds and thick darkness.
____   Only experts trained in theology are qualified to make sense out of Genesis 1.
____   Theistic evolution and progressive creation are attempts to reconcile the Bible with science.
____   Moses, though not a man of science in the modern sense, can be expected to have understood all aspects of the creation in Genesis 1.
____   Astronauts out in space wake up every morning to a cloudless dawn.

9. What is the problem with believing that creation was accomplished by means of evolutionary processes?

10. What is an allegory? Are there indications from the scriptures that this is how Genesis is to be read? What are some allegories in the Bible and how do you know they are allegorical?

11. Should the propositions "It is uncertain whether or not a creator exists" or "It is certain that a creator does not exist" require as much of a rational defense as "It is certain that a creator exists"?

12. What might be a suitable reply to the skeptic who says that Genesis 1 is a myth or fable?

13. In John 18, the Lord takes a stand for truth before Pilate at the cost of his life. What does this passage imply about the need for objectivity in what is believed?

Chapter 3
Questions:

1. The principle of uninterrupted uniformity ... (T / F)

____ was a deduction from a scientifically inspired eighteenth-century theology.

____ was first described by Charles Lyell.

____ was associated with early geology.

____ is compatible with the Genesis creation record.

____ considers supernatural phenomena.

____ was first explained as "The present is the key to the past."

____ is called "uniformitarianism" by geologists.

____ is the idea that natural processes were the same in the past as they are today.

____ became the basis for scientists to devise theories of natural history.

2. T / F

____ The heroic age of geology ran from 1830 to 1833.

____ Geology was revised to correlate better with Genesis.

____ Genesis was reinterpreted to reconcile it with geology.

____ An isotope is a molecule of high atomic mass.

____ The radio-isotope dating technique must be applied to sedimentary rocks.

____ The geologic column is dependent on an evolutionary history or faunal succession of index fossils.

____ The bottom layers, being the youngest, were deposited first.

____ A half-life is the time for exactly one half of the radioactivity to go away.

____ Lord Rutherford discovered radioactivity in 1905.

____ An important assumption in radio-isotope dating is that radioactive decay rates have been constant.

_____ The complete geologic column may only be found at a few locations on Earth.
_____ The initial concentration of the radio-isotope to be applied must be known in order to date a rock sample.
_____ An unstable radio-isotope has extra neutrons compared to the normally occurring stable element.
_____ A missing stratum is one example of what geologists call an unconformity.
_____ Until radio-isotope dating was perfected, geologists could only speculate about absolute ages.
_____ The geologic column is the foundation of uniformitarianism.
_____ Carbon-14 dating can only be performed on inorganic materials.
_____ Sets of facts can often be interpreted or construed in more than one manner.

3. Catastrophism favored which natural events or processes?
   a. Floods
   b. The formation of shallow seas
   c. Continental drift
   d. Earthquakes
   e. Volcanic eruptions
   f. Deposition or sedimentation
   g. Erosion

4. Uniformitarianism favored which natural events or processes?
   a. Floods
   b. The formation of shallow seas
   c. Continental drift
   d. Earthquakes
   e. Volcanic eruptions
   f. Deposition or sedimentation
   g. Erosion

5. Why did uniformitarianism prevail over catastrophism?
   a. It required more time and made the Earth older.
   b. It was promoted by the "father of modern geology."
   c. It was better at explaining numerous rock layers.
   d. It required only presently observed natural processes.
   e. It had more empirical support.
   f. Nonconforming evidence could be disregarded.
   g. All of the above

6. Why was the Genesis record of a worldwide flood eventually abandoned by naturalists?
  a. Naturalists assumed that each rock layer was formed by a separate event.
  b. The rate of sedimentation as observed today would take a long time to form the known sedimentary rock layers.
  c. Naturalists refrained from appealing to supernatural explanations.
  d. No one could figure out how one flood could cause so many layers of sedimentary rock.
  e. The types of fossils observed in the various rock layers were different.
  f. All of the above

7. Uniformitarianism is a deduction from which philosophy?
  a. Deism
  b. Theism
  c. Atheism
  d. Scientism
  e. Skepticism
  f. Evolutionism

8. The geologic column is …
  a. the basis of the geochronological time scale.
  b. ordered from oldest (bottom) to youngest (top).
  c. incompletely observed in type sections.
  d. stacked according to an ordering of index fossils on a global scale.
  e. dependent on an evolutionary time sequence of index fossils.
  f. not found in its entirety at any one locality on the globe.
  g. all of the above

9. Radio-isotope dating is …
  a. dependent on assumptions of initial isotope concentrations that are, in turn, dependent on the study of the formation that a sample was taken from.
  b. the means geologists use to determine absolute ages of rock formations.
  c. generally valid only when applied to igneous rocks.
  d. valid only if the true age of a sample is a significant fraction of the radio-isotope half-life.
  e. a method that was developed after the idea that the Earth was many millions of years old was already established.
  f. different from carbon dating in that it is applied to inorganic material.
  g. all of the above

10. Why is the formation of fossils incompatible with the slow accumulation of sediment posited by uniformitarian geology?

11. What does the presence of carbon-14 in coal deposits and diamonds likely indicate about their age?

12. Since neither uniformitarian geochronology nor the Genesis record of a globe-wide flood can be validated empirically, what must uniformitarian geologists rely upon and what must creationists rely upon to validate their favored theory on rock formation?

13. Is the principle of uninterrupted uniformity a conclusion or an assumption of natural history?

Chapter 4
Questions:

1. T / F

____ Biological evolution was a scientific theory before becoming a philosophy.

____ Charles Darwin published *Origin of Species* in 1809.

____ Charles Darwin invented the term *natural selection*.

____ Thomas Huxley invented the term *agnostic*.

____ "Inheritance of acquired characteristics" was an empirically based hypothesis.

____ Charles Darwin's evolutionary theory was dependent on uniformitarian geochronology.

____ Georges Cuvier linked the natural history of fossilized life forms with uniformitarian geochronology.

____ Charles Darwin's evolutionary theory was hatched in an ideological vacuum.

____ The industrial revolution had inspired an ideology of technological and social progress.

____ Ideas about developments in living things were influenced by technological and social progress in the 1800s.

____ Selective breeding of domestic animals and plants was considered a form of controlled evolution in the 1800s.

____ Charles Darwin's evolutionary theory was an extrapolation of his empirical findings.

____ Charles Darwin's *Origin of Species* was slow in raising theological and ethical controversy.

____ Natural selection is defined as branching lineages from common ancestry.

_____ The neo-Darwinian synthesis consisted of natural selection operating on random genetic mutations.

_____ Lamarck introduced the concept of the inheritance of acquired characteristics in 1809.

_____ Charles Darwin had first thought of the idea of natural selection in 1838.

_____ Alfred Russel Wallace devised an explanation of species development similar to Charles Darwin's.

_____ Michael Ruse addressed interactions between evolution as science and evolution as a philosophy of culture.

_____ Thomas Huxley was known as Charles Darwin's bull dog, due to his advocacy of natural selection.

_____ DNA was discovered in 1953 and completed the neo-Darwinian synthesis.

2. Evolution is …

    a. something that occurs on the Galapagos Islands.

    b. a scientific theory.

    c. a theory consisting of natural selection operating on random genetic mutations.

    d. a theory dependent on the extrapolation of observations.

    e. incremental change over long periods.

    f. an idea originated by Charles Darwin.

    g. generally regarded among scientists as the central organizing concept in biology.

    h. all of the above

3. Scientists were initially hesitant to accept Darwin's evolutionary theory for which of the following reasons?

    a. Evolution had fared poorly with scientists in the past.

    b. It raised ethical and theological questions.

    c. It did not rely on supernatural causes.

    d. They already had a sufficient number of theories.

    e. All of the above

4. Scientists began to accept Darwin's evolutionary theory for which of the following reasons?

    a. The detail and systematic nature of his observations.

    b. It gave scientists a focus for research.

    c. It did not rely on supernatural causes.

d. They had grown tired of the idea of creation.

e. All of the above

5. Darwin's evolutionary theory consisted of which two of the following concepts?

a. Natural selection

b. Inheritance of acquired characteristics

c. Orthogenesis

d. Neo-Darwinian synthesis

e. Branching lineages from common ancestry

f. Environmental pressure

g. Random genetic mutation

6. Natural selection fell from favor in the late nineteenth century when …

a. theistic evolution took over.

b. Charles Darwin died.

c. Thomas Huxley decided it was not true.

d. scientists wanted a mechanism that could explain variations.

e. all of the above

7. Who was able to leverage Darwin's theory to exclude supernatural causes from science?

a. Michael Ruse

b. Charles Darwin

c. Theodosius Dobzhansky

d. Thomas Huxley

e. Julian Huxley

f. Gregor Mendel

8. The neo-Darwinian synthesis combined which two of the following concepts?

a. Secular humanism

b. DNA

c. Random genetic mutations

d. Inheritance of acquired characteristics

e. Genetics

f. Natural selection

9. What cultural transition began to appear in the late nineteenth century as a consequence of Darwin's theory?

    a. Science displaced philosophy in Western culture.
    b. Science displaced theology in explaining the origin of mankind.
    c. Genetic mutations replaced inheritance of acquired characteristics.
    d. Secular humanism replaced theistic religion.
    e. Darwinism was replaced by neo-Darwinism.
    f. All of the above

10. Secular humanism …

    a. was an attempt to consciously rethink the philosophy of culture.
    b. was justified by evolutionary science.
    c. was a proposal for reason to replace religion as the grounds of ethical behavior.
    d. was officially proposed in 1933.
    e. entered public education in the early 1960s upon the reintroduction of evolutionary science.
    f. is based on an assumption that science is the only means to obtain knowledge.
    g. all of the above

11. Why did Thomas Huxley invent the term *agnostic*?

12. The transformation of knowledge, which began during the scientific revolution in the seventeenth century and was completed by Darwin's theory, may be described as a transition from a _____ basis.

    a. natural to a supernatural
    b. philosophical to a scientific
    c. theological to a philosophical
    d. philosophical to a natural
    e. theological to a scientific

13. What aspects of biological evolutionary theory and its history indicate that it was and still is a philosophical idea?

14. What are two long-standing arguments against the veracity of biological evolution? How do you think it is possible that evolutionary theory can survive these?

15. What does the history of evolutionary theory illustrate about how non-scientific ideas can gain scientific credibility?

Chapter 5
Questions:

1. Which of the following was/were known in ancient Greece?
    a. Diameter of Earth
    b. Diameter of the moon
    c. Diameter of the sun
    d. Distance between Earth and the moon
    e. Distance between Earth and the sun
    f. Circumference of Earth's orbit around the sun
    g. All of the above

2. How was it known in the seventeenth century that the Earth revolves around the sun and not the sun around the Earth?
    a. Copernicus and Galileo both said so.
    b. The perception of motion was understood to be relative, explaining why the Earth's motion might not be felt.
    c. It was understood that objects are attracted to Earth because of gravitation, and not because the Earth is the center of the universe.
    d. The summer and winter solstices indicated a nonuniform trajectory of the sun that could not be explained by Newton.
    e. Archaic motion, required by the Earth-centered model, was not supported by laws of planetary motion and gravitation.
    f. No one could determine if the positions of nearby stars shifted against the background of other more remote stars.
    g. Kepler discovered that planetary orbits were elliptical.

3. What means confirmed this theory (heliocentricity) in the nineteenth century?

   a. Charles Messier differentiated between comets and nebulae.
   b. The Milky Way was finally identified as a spiral galaxy.
   c. Knowing the mass of the sun as compared to that of Earth.
   d. Successful parallax measurements of the distance to nearby stars.
   e. Planetary orbits could best be explained as ellipses rather than circles.
   f. All of the above

4. Johannes Kepler made use of the observations of planetary trajectories documented by whom?

   a. Tycho Brahe
   b. Galileo
   c. Edwin Hubble
   d. Neil Armstrong
   e. Nicolaus Copernicus
   f. No one; he used his own observations

5. Which three of the following make up Kepler's laws of planetary motion?

   a. All of the planets revolve around the two foci of their ellipses.
   b. All of the planets have elliptical orbits, with the sun stationed at one of two foci,
   c. The bipolar plane is formed by the intersection of two foci at the orbital extremes.
   d. The elliptical plane, when bisected by a straight line, forms two equal areas.
   e. A straight line between the planet and the sun covers equal amounts of area per unit time.
   f. The duration of planetary orbits is proportional to their mean radius from the sun.
   g. The duration of the orbital radius is proportional to the planet's proximity to the sun.

6. Claudius Ptolemy's planetary system was characterized by which of the following?

   a. Elliptical orbits
   b. Circular orbits
   c. Foci
   d. Epicycles
   e. Carousels
   f. All of the above

7. The baseline for measuring distances outside the solar system is what?

    a. The radius of the sun
    b. The radius of the Earth's orbit around the sun
    c. The diameter of the Earth's orbit around the sun
    d. The diameter of the Earth
    e. The distance between the Earth and Mars

8. Beyond the range of parallax, distance measurements rely on which of the following?

    a. The number of variations in the brightness of objects per unit of time
    b. The apparent brightness of objects in comparison to their absolute brightness
    c. The amount of gravitationally induced redshifting in the light spectra
    d. The distance to the Small Magellanic Cloud
    e. Interference by intergalactic matter
    f. All of the above

9. The contradiction between the laws of electromagnetics and the laws of motion was resolved by whom?

    a. Galileo
    b. Newton
    c. Hubble
    d. Einstein
    e. Lemaitre
    f. Slipher

10. A light-year is which of the following?

    a. An exceptionally bright year
    b. A new diet plan
    c. The distance light traverses through empty space in one year

11. Einstein said that the measured speed of light would be …

    a. much faster than could be calculated or measured.
    b. relative to the speed of the observer with respect to the source of light.
    c. independent of the speed of the observer with respect to the source of light.
    d. increased in accordance with the length of the distance traversed.
    e. quicker than the hand but not the eye.

12. What did Einstein say is/are absolute and universal?

    a. The laws of physics
    b. Time and distance
    c. The speed of light
    d. The force of gravity
    e. The uniformity of space and time
    f. All of the above

13. The use of Cepheid variables to estimate distances in space is dependent on what?

    a. Knowing their average apparent brightness as a function of the period of variation
    b. Knowing their absolute brightness in comparison to their apparent brightness
    c. Having accurate distance measurements to a representative sample
    d. The assumption that the periodicity-brightness relation applies throughout the universe
    e. All of the above

14. Redshifting means that the frequencies of visible light …

    a. decrease due to a light source moving toward Earth.
    b. increase due to a light source moving toward Earth.
    c. decrease due to a light source moving away.
    d. increase due to a light source moving away.

15. The realization that the universe is expanding was based on which of the following?

    a. Galileo's principle of relativity
    b. Einstein's theory of relativity
    c. Hubble's redshift and distance measurements
    d. Gamow's big bang theory
    e. Lemaitre's interpretation of relativity theory
    f. The principle of uniformity
    g. The cosmological constant
    h. All of the above

16. Astronomers estimate the age of the universe by doing what?
   a. Calculating the average of distances to remote galaxies and dividing by the average speed of remote galaxies
   b. Considering the rate of nuclear fusion and the remaining hydrogen and helium in the oldest stars
   c. Considering the time light takes to get from the most remote objects seen in space to arrive on Earth
   d. All of the above

17. Observation of the universe as it exists today is theoretically limited by what?
   a. Telescope technology
   b. The event horizon
   c. Non-Euclidean geometry of space
   d. The finite speed of light
   e. Dark matter
   f. Space dust and debris
   g. All of the above

18. What three theories about the universe have been overthrown by science?
   a. Big bang theory
   b. Geocentrism
   c. Ether theory
   d. Relativity theory
   e. Dark matter theory
   f. Stationary universe
   g. Heliocentrism

19. T / F
   ____ Parallax is a triangularization technique that astronomers use to measure distances to objects in space.
   ____ Copernicus based his theory not on observations, but on mathematical reasonableness.
   ____ Einstein based his theory not on observations, but on consistency in the laws of physics.
   ____ Cepheid variables remain the best means to estimate distances to neighboring galaxies.
   ____ Spectroscopy is the analysis of the composition of sources of light, such as stars, from their light spectrums.

____ A correlation between physical reality and mathematical logic was taken for granted in the seventeenth century.

____ Parallax measurements are dependent on the precision attainable in measuring very small angles.

____ Parallax is just one among several accurate and precise means of measuring distances to objects in space.

____ Direct parallax measurements of Cepheid variables have not been attained to date.

____ Einstein resolved a contradiction between laws of motion and those governing electricity and magnetism.

____ Blueshifted spectral lines in a light source indicate the source has a receding velocity.

____ The identity of nebulae was uncertain until Edwin Hubble began to estimate their distances in the 1920s.

____ Edwin Hubble was the first astronomer to use spectroscopy to determine the radial velocity of nebulae.

____ Einstein's cosmological constant was a means for relativity theory to permit a stationary universe to exist.

____ The big bang theory of the universe is based solely on a deduction involving the redshifting of light from remote galaxies.

____ The microwave background radiation is thought to be a remnant of the initial "bang" that formed the universe.

____ Stars typically produce light by means of nuclear fusion reactions wherein helium is converted to hydrogen.

____ Arthur Eddington obtained the first evidence that Einstein's theory of gravity was correct.

____ Dark matter is a hypothesis used to explain peculiarities in the rotation of spiral galaxies.

____ Events in remote space, such as supernovas, support the light-in-transit argument for creation.

____ Russell Humphreys devised a cosmological theory that can be correlated with Genesis 1 chronology.

____ The big bang theory takes into account relativity effects on elapsed time.

____ An event horizon is the boundary where gravity becomes strong enough to bend light back on itself.

____ Agreement between observation and mathematically based theory is a persuasive argument for the validity of any scientific conclusion.

20. What is the one overlooked factor in approaches to estimating the age of the universe that may yet render all attempts pointless?

21. What facts should moderate scientists' confidence that they have solved the mystery of the universe's formation and age?

Chapter 6
Questions:

1. What four qualifications are characteristic of successful scientific theories?

    1.

    2.

    3.

    4.

2. T / F

    ____   A successful scientific theory is always a true scientific theory.

    ____   Science may be described as a procedure to generate and then confirm or disprove hypotheses about cause-and-effect operations in nature.

    ____   Successful scientific theories bring coherence to assorted facts about nature.

    ____   Supernatural causes are not permitted in science as practiced today because, if accepted, they would end further inquiry into natural causes.

    ____   Scientific theories that are confirmed by experimental results and observations must be true.

3. _____ about operations of cause and effect are either confirmed or disproved by reference to empirical findings.

4. _____ are generated when _____ become widely accepted after having undergone rigorous peer review.

5. _____ are supposed to always have tentative statuses in that future discoveries may disprove them.

6. What are the two means that scientists use to confirm theories about processes in nature occurring in the present? What makes these means less applicable to processes in nature thought to have occurred in the unobserved past?

7. If you were to attempt to put together a 1,000-piece jigsaw puzzle without the aid of the picture on the box, would you be able to do it? How important is it to have the box as a guide? What might this illustrate about the interpretation of facts applicable to natural history, such as rock strata and fossilized remains?

8. Why must scientific theories always be regarded as tentative?

9. Why are theories confirmed by experimental results and observations not necessarily true?

10. What is the fundamental exclusionary stipulation of all modern scientific theory building?

11. What is the preferred assumption regarding scientific inquiry into the unobserved past?

12. What is the mistake in reasoning involved when, because of scientific theories on origin, natural causes are concluded to be sufficient to explain everything that exists?

13. Do you think that inquiry into the origin of the universe, Earth, and living things is a legitimate subject for science? Why or why not?

14. What is the mistake in reasoning involved when, because of scientific theories on origin, human existence is concluded to be without purpose? (Hint: Do you remember how purpose and design were excluded from scientific inquiry?)

15. What are some reasons that explain why scientific inquiry takes place at all? Can these reasons be explained scientifically?

Chapter 7
Questions:

1. What became the settlement between science and theology in the late nineteenth century?

    a. Evolutionary consensus

    b. Humanitarian atheism

    c. Theistic evolution

    d. Cooperative modernism

    e. Theological orthogenesis

    f. Scientology

2. Why was Darwin's natural selection at first discarded by scientists?

    a. Thomas Huxley opposed it

    b. It suggested no cause of variations

    c. Genes had not yet been discovered

    d. It was an idea ahead of its time

    e. Darwin had changed his mind

    f. All of the above

3. Higher or historical criticism was …

    a. an attempt to discount reports of miracles in the Bible.

    b. an idea on which new interpretations of scripture were based.

    c. accepted by some Protestant clergy.

    d. predicated on the idea that the scriptures were not divinely inspired.

    e. partly responsible for liberalization within Christianity.

    f. all of the above

4. What was lacking in the attempt to find an acceptable middle ground between science and religion in theistic evolution?

    a. There was nothing lacking
    b. The theistic aspect
    c. The evolutionary aspect
    d. Assent by a majority of scientists and theologians
    e. Any evidence that the proposition was true

5. What developments led to the 1925 Scopes trial?

    a. The emergence of secondary education
    b. Clarence Darrow's campaign for academic freedom
    c. Social Darwinism
    d. Fundamentalist activism
    e. William Jennings Bryan's campaign against teaching evolution
    f. The return of natural selection to evolutionary theory
    g. All of the above

6. The Scopes trial was an attempt to do what?

    a. Outlaw the teaching of evolution
    b. Punish John Scopes for teaching evolution
    c. Elevate the status of monkeys in society
    d. Challenge the constitution
    e. Challenge the constitutionality of the Butler Act
    f. Generate more business in Dayton
    g. Suppress academic freedom

7. The Darrow-Bryan interrogatory …

    a. affected public perceptions about the credibility of the Bible.
    b. was exploited by reporters and, later, historical writers in the 1930s.
    c. took place during the Scopes trial when the defense called Bryan to the witness stand.
    d. was a surprise strategy by Darrow to ridicule literal interpretation of the Bible.
    e. was extraneous to the necessary procedures of the court trial.
    f. all of the above

8. Why was the Scopes verdict not appealed to the US Supreme Court?

    a. Clarence Darrow was held in contempt of court.
    b. William Jennings Bryan had died.
    c. John Scopes quit his teaching job.
    d. The statute of limitations had passed.
    e. There was a technicality involving the fine that the Tennessee judges used to invalidate the jury's verdict.
    f. The ACLU had bigger fish to fry.

9. The legacy of the Scopes trial in the public mind has been distorted by which of the following?

    a. The play *Inherit the Wind*
    b. Myths perpetuated by some historians
    c. H. L. Mencken
    d. The press's continuing ridicule of fundamentalist beliefs
    e. All of the above
    f. None; it was not distorted

10. Aside from opportunistic exploitation by the media, can you explain why the Scopes trial, a court proceeding over a very low-level crime (if you can even call it that) is sometimes referred to as the trial of the century?

11. The neo-Darwinian synthesis …

    a. was completed in the 1950s.
    b. consisted of random gene mutation and natural selection by environment and competition between species.
    c. was declared a triumph of science and the centerpiece of biology.
    d. increased confidence among scientists that they had mastered biology.
    e. suggested that life had purpose.
    f. all of the above

12. What factors, aside from the neo-Darwinian synthesis, led to the reentry of evolution in science classes in the 1960s?

    a. The Cold War and associated concern over the state of science education
    b. The influence of *Inherit the Wind* on popular opinion
    c. The increasing secularization of Western culture
    d. The appeal of autonomous self-determination
    e. The Soviet's first successful satellite launch
    f. All of the above

13. Challenges to the teaching of creation in science classes have appealed to …

    a. the First Amendment's Establishment Clause.
    b. the right of state legislatures to regulate what is taught in public schools.
    c. common sense.
    d. the separation of church and state.
    e. the idea that evolution is science and creation is religion.
    f. all of the above

14. *Epperson v. Arkansas* was a court case involving which of the following?

    a. The State of Tennessee
    b. Creation science
    c. Equal time for creation and evolution in science classes
    d. A statute prohibiting the teaching of evolution
    e. The US Supreme Court
    f. Intelligent design

15. *McLean v. Arkansas* was a court case involving which of the following?

    a. The State of Tennessee
    b. Creation science
    c. Equal time for creation and evolution in science classes
    d. A statute prohibiting the teaching of evolution
    e. The US Supreme Court
    f. Intelligent design

16. *Edwards v. Aguillard* was a court case involving which of the following?

    a. The State of Tennessee

    b. Creation science

    c. Equal time for creation and evolution in science classes

    d. A statute prohibiting the teaching of evolution

    e. The US Supreme Court

    f. Intelligent design

17. *Kitzmiller v. Dover* was a court case involving which of the following?

    a. The State of Tennessee

    b. Creation science

    c. Equal time for creation and evolution in science classes

    d. A statute prohibiting the teaching of evolution

    e. The US Supreme Court

    f. Intelligent design

18. The modern practice of science is universally predicated on what?

    a. Experimental evidence

    b. The exclusion of supernatural causes

    c. Confirmed hypotheses

    d. A sun-centered solar system

    e. Non-intelligent design

    f. All of the above

19. Creation science …

    a. was employed in an initiative to balance the teaching of evolution with an equal emphasis on creation.

    b. advanced evidence that supported the literal six-day creation and flood of Genesis.

    c. was rejected by the courts as an attempt to insert religion into public-school science curriculums.

    d. was not intended by its originators to be imposed as a legal solution on public-school science curriculums.

    e. all of the above

20. The two main arguments of intelligent design involve …

    a. deductions from naturalistic philosophy.
    b. the information content of genes.
    c. the five-part mousetrap.
    d. mousetraps with a reduced number of parts.
    e. the irreducible complexity of living organisms.
    f. the increasing complexity of living organisms.

21. The exclusion of supernatural causes in science is …

    a. a great idea.
    b. a lousy idea.
    c. a chief conclusion of the scientific method.
    d. a stipulation in the practice of modern science.
    e. an unintended consequence of the partitioning of thought.
    f. a legacy of the scientific revolution.

22. How can scientists get away with claiming to have explained the origin of the universe and all living things, yet at the same time admit that scientific inquiry is limited to the natural world?

23. Is all existence natural because that is all science can deal with, or has science improperly claimed the subject of origin?

24. Confusion surrounding the evolution-creation controversy may be attributed to which of these?

    a. The separation of church and state
    b. Incremental changes of opinion over long periods of time
    c. A premise put forward as if it were a conclusion
    d. The wide variety of scientific origin theories
    e. The lack of intervention by God
    f. All of the above

25. Why have creation science and intelligent design fared so poorly in the courts?

    a. Evolution is viewed as science; creation science and intelligent design are viewed as religion.

    b. Theistic evolution makes it appear that evolution does not generally contradict religion.

    c. A lack of solutions to the origin problem besides evolution that can seem plausibly naturalistic.

    d. They are both supernatural propositions and thus not permissible in science.

    e. The Establishment Clause of the First Amendment prohibits them.

    f. All of the above

26. How do we know that evolution is primarily a philosophical idea or ideology? As you answer this question, consider the following:

- Whether the idea or the findings used in support of it came first.
- How the direction of change in living things is always portrayed.
- The extent to which thought in subject areas outside of science is influenced.
- What must be done with present observations to be able to support it.
- Assumptions that have covered for a lack of empirical evidence.

27. In your view, were the various laws and other initiatives that first prohibited evolution in the public schools, and then later attempted to introduce creation science and intelligent design, constitutional or unconstitutional? Why?

28. If creation is fact and evolutionism is only an ideology or thought system based on assumptions that can be traced back to the scientific revolution, why has it been so difficult to establish it as a fact in public policy?

Chapter 8
Questions:

1. Even though the terminology used in describing cause-and-effect and rational inference is often the same, that should not obscure the distinction. Identify whether the word *because* in the following sentences describes a cause-and-effect (CE) relationship or a rational inference (RI) being made on cause-and-effect grounds:

____ The dog is clean because he was washed.

____ The dog must have been washed because he is clean.

____ The car is gone because she drove it to the store.

____ She must be at the store because the car is gone.

____ Her hands are cold because she forgot her mittens.

____ She must have forgotten her mittens because her hands are cold.

____ It will get dark soon because the sun is setting.

____ The sun must be setting because it is getting dark.

____ The phone must be disconnected because it has not been ringing.

____ The phone has not been ringing because it is disconnected.

____ The dog must be loose because we do not see any ducks.

____ We will not see any ducks because the dog is loose.

____ I thought you left your shoes outside because they are all wet.

____ We think they had car trouble because they are late.

____ They must have thought they were in danger because they went inside.

____ They went inside because they thought they were in danger.

____ Your shoes are all wet because you left them outside.

____ Your shoes will get wet because you left them outside.

____ They are late because they had car trouble.

____ It is getting dark because the sun is setting.

____ We do not see any ducks because the dog is loose.

Notice what rational inference does that sets it apart: given an effect only, its cause may be inferred (an explanatory inference), or given a cause only, an effect may be inferred (a predictive inference). Conversely, when a cause-and-effect relationship is described, both the cause and the effect are already known. In one case, there is a potential for new knowledge beyond experience; in the other there is not.

2. Objective thinking is …

    a. the result of sensory perception.
    b. the result of laws of nature.
    c. subject to laws of nature.
    d. always true.
    e. the result of laws of logic.
    f. subject to laws of logic.
    g. dependent on natural selection.
    h. subject to sensory perception.
    i. the mental exercise of discerning truth.
    j. the result of accumulated experience.
    k. all of the above

3. T / F

____ Knowledge of things or events beyond direct human experience must be inferred.

____ The mind is a product of a long sequence of cause-and-effect processes in nature.

____ Limited observations of processes in nature are sufficient to know about natural laws.

____ There is evidence to suggest that the brain and the mind have different origins.

____ Language is a process of cause and effect that conveys objective meaning.

____ The meaning of words becomes subject to the hearer or reader once they are received.

____ Attributing someone's thoughts to natural causes is a means used to establish the validity of those thoughts.

____ As soon as thought becomes subject to some cause-and-effect process, it ceases to be objective.

___ An indication that words have objective meaning is the common assent to dictionary definitions.

___ The mind comprehends laws of nature because laws of nature produced the mind.

___ Objective thought was forged against a survival criterion and not a truth criterion.

___ Scientific knowledge involves insight into principles and factors underlying observed patterns in nature.

4. Thought is a different process than cause-and-effect because …

    a. scientists cannot explain how thought occurs.

    b. thought is always *about* something.

    c. thought may be true or false.

    d. thought is subject to laws of logic, not laws of nature.

    e. thought can reach beyond sensory experience.

    f. thought is inevitable.

    g. all of the above

5. The law of noncontradiction allows that …

    a. there is a rational order to nature.

    b. objective thinking is true.

    c. laws of nature are universal.

    d. laws of logic result from natural order.

    e. self-consistent propositions may be true.

    f. natural order results from laws of logic.

    g. truth might not exist.

6. How do we know that words convey objective meaning?

7. If, as the law of noncontradiction implies, truth exists, then …

    a. assertions about what is true are credible.

    b. there is a subset of thought that is solely dependent for its existence on its being true.

    c. truth must be an effect of natural causes because inferences are not always reliable.

d. there is a universal rationality not subject to natural processes that rules thought.

e. truth must be an effect of natural causes because natural causes are the only causes that exist.

f. truth has arisen as the ultimate outcome of a process of natural selection.

g. I'm in serious trouble.

8. Must the validity of rational thinking be assumed at the start of any inquiry into truth?

9. Why do attempts to give an account of rational thought from cause-and-effect operations in nature fail?

a. Cause-and-effect events do not always conform to rationally known laws of nature.

b. They do not fail; knowledge of laws of nature could be an effect of the operations of laws of nature.

c. Cause-and-effect operations in nature cannot produce knowledge beyond experience.

d. An account of rational thinking cannot be given without using rational thinking to do it.

e. The validity of rational thinking must be assumed up front; after that, cause and effect can add no further validity.

10. Objectivity in science depends most fundamentally on which one of the following?

a. Laws of nature

b. The scientific method

c. Cause-and-effect processes

d. Laws of logic

e. Observation and reason

f. A rational order to which nature is subject and which the mind comprehends

11. What does the universal acceptance of the validity of rational thinking and the absurdity of attributing it to natural causes suggest about how it arises?

Chapter 9
Questions:

1. Philosophy may be described as a quest to ...

    a. understand the meaning of existence.
    b. expose hidden assumptions in claims of truth.
    c. discover coherency and unity in diversity.
    d. challenge questionable presuppositions.
    e. answer seemingly unanswerable questions.
    f. all of the above

2. The big problem with philosophy is in ...

    a. questionable premises.
    b. diminishing premises.
    c. disappearing premises.
    d. insufficient premises.
    e. inadvertent premises.
    f. universal premises.

3. What four philosophical questions has everyone some type of answer to, evident by the way they live, whether they consciously admit to having answers or not?

    1.
    2.
    3.
    4.

4. How does the fact that the first of the above four questions is assigned to science and the rest to religion expose the incoherence of the present-day cultural solution?

5. The ability to know anything beyond direct experience depends on …
    a. reasonable skepticism.
    b. presuppositions.
    c. sensory perception.
    d. adequate premises.
    e. rational inference.
    f. adequate evidence.
    g. all of the above

6. In philosophy, arguments are what show how _____ are reached from facts and premises.
    a. presuppositions
    b. perceptions
    c. inferences
    d. predispositions
    e. conclusions
    f. assumptions
    g. observations

7. Premises, presuppositions, presumptions, and assumptions are _____ terms for the starting points of arguments.
    a. standard
    b. synonymous
    c. appropriate
    d. principal
    e. philosophical
    f. argumentative
    g. historical

8. Philosophy has had a history of attempting to remove _____ as a premise.

9. Momentary sensory perceptions can be augmented by what two sources?

    1.

    2.

10. The practice of science is predicated on what three assumptions?

    1.

    2.

    3.

11. T / F

____ Successful scientific theories and successful philosophies bring coherence to assorted facts or experiences.

____ The coherence of successful scientific theories and philosophies means that they are true.

____ Characteristic of suspect scientific theories or philosophies is that they are built on assumptions or presuppositions.

____ Scientific theories built on assumptions that cannot be empirically verified are false.

____ A philosophy built on a presupposition that does not satisfy its own criterion for truth is false.

12. What distinguishes the methodology of naturalism from the philosophy of naturalism is that the methodology of naturalism excludes _____ causes in order to investigate natural causes, whereas philosophical naturalism _____ that natural causes are all there is.

    a. relevant / concludes
    b. supernatural / concludes
    c. effective / concludes
    d. effective / assumes
    e. supernatural / assumes
    f. practical / assumes
    g. supernatural / denies

13. T / F

____ Views regarding good and evil depend on a prioritization of values.

____ Views regarding good and evil become relative or subjective when a transcendent source of values is denied.

_____ Among the ramifications of philosophical naturalism is that there is no ethical or moral compass outside of individuals.

_____ A circular argument presupposes the point being argued.

_____ Because rationality must be assumed, there are no independent paths to verify the means by which knowledge is acquired.

_____ Philosophical naturalism is self-defeating in that it is put forward as an objective truth, but an objective truth cannot arise from natural causes.

14. Postmodernism's denial of truth and meaning proceeds from philosophical naturalism because …

a. postmodernism denies scientific knowledge.

b. naturalism assumes that nature is all that exists.

c. naturalism is a self-defeating premise.

d. postmodernism is a self-defeating premise.

e. naturalism provides no basis for truth or for meaning.

f. naturalism is a circular argument leading to postmodernism.

15. When presented with a scientific exhibit or a report of a scientific discovery, asking whether naturalism as a m_____ or naturalism as a p_____ is assumed helps clarify underlying presuppositions.

16. Why can questionable presuppositions or premises of a universal character not generally be overthrown by rational argument?

a. Rational arguments are ineffective.

b. Rational arguments are built on top of them.

c. Evidence can never be complete.

d. Other factors are often involved in choosing them.

e. Circular arguments cannot be overthrown by other circular arguments.

f. All of the above

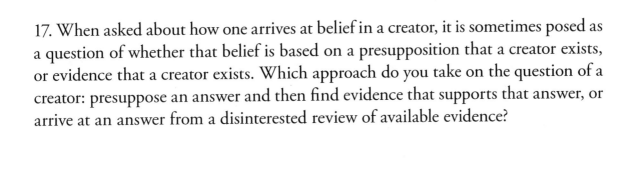

17. When asked about how one arrives at belief in a creator, it is sometimes posed as a question of whether that belief is based on a presupposition that a creator exists, or evidence that a creator exists. Which approach do you take on the question of a creator: presuppose an answer and then find evidence that supports that answer, or arrive at an answer from a disinterested review of available evidence?

Does it matter?

How many actually undertake a disinterested review of available evidence?

Is it even possible to undertake a disinterested review in the strict sense of the term *disinterested*?

What would become of arguments should the Creator show up in person?

[Of course, the question should be returned: Is non-belief in a creator a presupposition with some evidence clattering along behind, or is it arrived at by a disinterested review?]

18. Why are presuppositions such a big problem for philosophy and why has philosophy consistently failed to arrive at satisfactory answers to the questions it deals with?

Chapter 10
Questions:

1. How can philosophical naturalists get away with never mistaking manmade objects as effects of natural causes, while at the same time regarding human beings as an effect of natural causes?

2. Given the barriers thrown up against intelligent design by the scientific community, how might the case for design be made?

3. If _____ values are removed, ideas about right and wrong, and good and evil, become subjective preferences of individuals.

   a. evolutionary
   b. postmodern
   c. transcendent
   d. philosophical
   e. collective
   f. deterministic

4. How must naturalists account for the existence of ideas about right and good? Are there objective values hidden within these explanations? If so, what are they?

5. What must be the logical conclusion about design and value in the world if the presence of design and value in theories of origin is denied?

How are design and value apparent in, say, a scientific experiment?
How is it possible for design and senses of value to exist?

6. Can socially or biologically conditioned responses supply insight into or resolve questions involving the truth of a belief or the rightness of a public policy?

If not, how can such questions be resolved?

7. The following two statements might be termed the twin pillars of postmodern "truth." What is the problem with each?

There are no universals.
No proposition ever corresponds to reality.

8. Is objective knowledge possible if ultimate reality is strictly naturalistic and thus ruled by cause-and-effect? Why or why not?

9. What then is the problem with the "truth" of naturalism, that there is nothing that is not from or a part of nature?

10. What do heated controversies and debates over what constitutes right and wrong really indicate?

    a. Ideas about right and wrong are never absolute.
    b. Everyone should be able to decide what is right for them.
    c. Conflicts over the prioritization of values exist.
    d. People are prone to anger.
    e. This question cannot be answered with certainty.

11. The argument from evil, which is an attempt to disprove the existence of God, carries what hidden assumption along with it? What is it about this assumption that turns the argument back on itself?

12. What might one begin to say to expose the hidden assumption, stand the argument on its head, and hand it back to the skeptic?

    a. Are you saying that the existence of evil is objectively true?
    b. How can you be indignant over what amounts to effects of natural causes?
    c. Are you expecting me to agree with your views on good and evil?
    d. Any of the above

13. The sense of rational order by which all knowledge beyond experience, including scientific knowledge, is reached is …

    a. inferred from experience.
    b. assumed.

14. What answer to the question of the meaning of existence does the logic of a scientific approach to human origin inevitably produce?

15. How do common everyday experiences involving meaning, morality, and destiny suggest that there is objectivity to be found?

16. Besides internal consistency, what other test of the authenticity of a philosophical presupposition is presented in this chapter?

17. What is an effective way to discredit someone's belief or assertion that does not involve identifying faulty premises or reasoning? Why do causal grounds leave no room to rationally establish a belief or assertion?

18. How might a perception of ambiguity regarding the existence of the Creator adversely affect the freedoms and rights of individuals?

Chapter 11
Questions:

1. What characterizes the state of affairs on Earth regarding beliefs about that which is beyond nature, and how has God leveraged that state of affairs to authenticate his message to mankind?

2. What point about the Old and New testaments begins the line of reasoning that leads to the conclusion of their divine authorship? (For an introductory study on the specifics of the qualifications of divine authorship, see Jay Wilson, *Proof that the Bible is the Word of God*, 11th Hour Press, 2002)

3. What does the hostility of the Jews in the first century toward the establishment of Christianity on the authority of the Old Testament indicate about the authorship of the Old Testament?

4. *Judeo-Christian* is a term often tossed around and it may be thus assumed that Jewish and Christian beliefs are compatible. Why is this a false assumption?

5. What does Paul say he is on trial for in Acts 23, 24, and 26, and how does what he said bear on the answer to Question 4?

6. What are the four lines of attack against the resurrection? Why are they each more absurd than the supposed absurdity they attempt to dismiss?

7. Why is the resurrection such an audacious and yet central move by God?

8. How does the Bible comfort fears and sustain hopes regarding mortality in a way that is superior to any religion?

9. What three tests have been applied to this point in the book to validate a religious or philosophical premise?
    a. Whether or not it affirms design in nature
    b. Whether or not it can stand by its own logic
    c. Whether or not there are hidden presuppositions
    d. Whether or not there are qualified empirical credentials
    e. Whether or not it is backed by science
    f. Whether or not it is relevant to meaning and morality
    g. Whether or not it is a forcefully imposed belief

10. If there was no death, would there still be a need for religion or revelation?

11. Is depending on nature to argue for the existence of a creator a strategic error? If so, how might that be avoided?

Chapter 12
Questions:

1. The sense of the term *faith* that has been assumed by many scientists and philosophers is which of the following?

    a. Confidence or trust in another person

    b. Unquestioned belief that requires neither proof nor evidence

    c. The assurance of things hoped for and the conviction of things not seen

    d. Belief in superstition and myth

    e. The ability to continue to believe in an unseen reality already verified as true

    f. A substitute for reason

2. The sense of the term *faith* as it is described or assumed in the scriptures is which of the following?

    a. Confidence or trust in another person

    b. Unquestioned belief that requires neither proof nor evidence

    c. The assurance of things hoped for and the conviction of things not seen

    d. Belief in superstition and myth

    e. The ability to continue to believe in an unseen reality already verified as true

    f. A substitute for reason

3. What is the one thing the scriptures state is contrasted with faith? What is faith never contrasted with, though often thought to be?

4. What is one way in which the apostles and writers of the New Testament show that God appeals to reason?

5. What is one way in which the apostles and writers of the New Testament show that God also appeals to faith?

6. Faith is often thought to be a substitute for _____. What is the real role of faith?

7. What can happen when you have faith without reason? Are moral judgments reasoned judgments? If so, reasoned from what premise?

8. What can happen when you have reason, but no faith? What can overtake reason if faith is absent?

9. Faith is a venture more in response to _____ __ _____ than to _____ _____.

10. Must faith always follow reason? When does it not follow reason? Is there ever an intuitive quality about it?

11. What is often claimed as the reason for skepticism toward the unseen reality of the Creator? What may likely be the real reason(s)?

12. Why does God not just provide supernatural appearances or miraculous signs from time to time to sustain belief in Him?

Answer
Key

Chapter 1

1. T, F, F, F, T, F, F, T, F, T, F, F, F, T, T, T, F, T, F, T, F

2. f

3. a, b, c, d, e

4. a

5. e (a is okay too)

6. a

7. e

8. e

9. b, c, e, g

10. a, c

11. The rational approach

12. Composition cause, formal cause, and final cause / Mechanical philosophy

13. Empirical verification and mathematical description

14. The supernatural. It was an aid; the supernatural was regarded as rational by the most successful of the natural philosophers.

15. (1) The collecting of facts by observation and experimentation. (2) Generalization or induction from observed facts to unifying concepts, theories, or natural laws. It subjected theories about nature to rigorous verification.

16. It produced the theology (or philosophy) of deism, the theology that says that since the creation, laws of nature alone have ruled.

17. Joshua 10:13. The sun is described as standing still, but it is a matter of perspective; we still say that the sun rises and sets even when we well know that the earth rotates.

18. Uninterrupted uniformity of natural processes since the initial creation (a consequence of, or a deduction from, deistic theology or philosophy).

19. The idea of rational order in nature. The lack of acceptance of that idea in the Islamic Empire and China led to collapse of scientific progress, even though both were more scientifically advanced earlier on.

20. An up-front decision to include or not include the consideration of design.

Chapter 2

1. F, F, T, T, F

2. They appeared solely as attempts to reconcile Genesis with recently devised uniformitarian geologic theory.

3. It is helpful in determining whether the meaning of *day* is one rotation of the earth or some other time period, and also in clarifying Genesis 1:2. The meaning of *day* may be established with more certainty as one rotation of the earth by considering certain statements of fact in Psalm 104, Job 38, and 2 Peter 3 that clarify that seas covered the earth at the beginning of creation and that the darkness was due to a thick cloud cover. It then becomes clear from Genesis 1:2 that the Spirit of God was "moving over the [sur]face of the" seas. The evenings and mornings associated with each creation day are then understood as sunsets and sunrises.

4. Old Testament poetry often pairs phrases that convey the same or similar ideas with different words.

5. Because He (God) saw that what was created was good within the boundaries (evening and morning) demarking each day.

6. Most certainly because of the connection with the Sabbath day in Exodus 20:11 and because Moses interpreted the Sabbath-day commandment on the basis that the days of creation were twenty-four-hour days. Context needs to be studied to determine the meaning of a word that has more than one sense or meaning.

7. The Sabbath-day commandment is the only explanation for it.

8. T, F, F, T, F, T, T, F

9. There is nothing to draw upon from the book of Genesis that indicates anything was created this way. Genesis 1 indicates that things were created rather abruptly (within the confines of a day).

10. A story with a symbolic meaning, such as a parable. There are no indications from the Genesis 1 text or from the Bible as a whole that this chapter is to be

taken as allegorical. Some allegories are identified as allegories, such as those in Ezekiel chapter 17 and Galatians chapter 4. Other allegories are recognized as allegories from their context, such as those in Isaiah 5, Ezekiel 15 and 16, or any of the parables in the New Testament.

11. (A discussion question)

12. The idea that it is a myth or fable seems to have arisen in the manner of a myth or fable (that is, over time) more than the idea that it is true. (Note that this charge is much easier to raise than to dispel. The best approach is probably to first disarm its insinuating force, if possible, by pulling it down to the level of a proposition to be evaluated on the basis of whether or not it also is a myth or fable.) The credibility of this text is ultimately established on the consistency of Old and New Testament references to it as well as on the credibility of the Bible as a whole.

13. (A discussion question)

Chapter 3

1. T, F, T, F, F, T, T, T, T
2. F, F, T, F, F, T, F, T, F, T, F, T, T, T, T, F, F, T
3. a, d, e
4. b, f, g (continental drift was not discovered until later)
5. d
6. a, c, d, e
7. a
8. g
9. g
10. As remains of living things or their imprints, they deteriorate rapidly if not preserved by immediate burial.
11. They are less than 50,000 years old, the maximum age range of C-14 dating.
12. Uniformitarian geologists must rely on the principle of uniformity of natural processes in the unobserved past (or the irrelevance of supernatural causes); creationists must rely on the authority of the creation time scale and flood accounts in the Bible.
13. An assumption

Chapter 4

1. F, F, T, T, F, T, T, F, T, T, F, T, F, F, T, T, T, T, T, F, T
2. b, c, d, e, g
3. a, b
4. a, b, c
5. a, e
6. d
7. d
8. c, f
9. b
10. g
11. To express the inability of scientific inquiry to determine whether or not God exists or anything about God.
12. e
13. It began as a philosophical idea that was not accepted as scientific for at least fifty years. During the nineteenth century, the idea of natural progress grew in conjunction with a cultural ideology of technological and social progress. It is always implicitly presented in terms of progress, and never regress. It also depends on uniformitarianism, which is itself a philosophical idea.
14. (1) There is a total lack of empirical evidence that living things can arise from nonliving matter. (2) There is a total lack of empirical evidence that new species can arise from existing ones. It is, at its root, a philosophy or ideology and thus not entirely dependent on evidence.
15. They can gain credibility merely by association if they are applied to explain scientific facts.

Chapter 5

1. a, b, d
2. b, c, e
3. d
4. a
5. b, e, f
6. b, d
7. c
8. b
9. d

10. c

11. c

12. a

13. e

14. c

15. b, c

16. d

17. d

18. b, c, f

19. T, T, T, T, T, F, T, F, T, T, F, T, F, T, F, T, F, T, T, F, T, F, T, T

20. Relativity of time and space.

21. (A discussion question); possible answers include (1) the still unverified dark matter proposed to account for certain observations on galaxy rotation; (2) an inability to determine what percentage of the universe is observable; (3) the limitations on observation imposed by the finite speed of light; (4) the exclusion of relativity effects on time and distance in age calculations; (5) the problem of first causation; (6) the existence of minds that demand explanations of the universe.

Chapter 6

1. Explains scientific facts; testable; makes successful predictions; widely accepted by scientists.

2. F, T, T, T, F

3. Hypotheses

4. Theories / hypotheses

5. Theories

6. Observation and experiment. Obviously, there is no observation of events in the unobserved past; experiments that attempt to replicate hypothesized events in natural history are often impractical or impossible.

7. (A discussion question); the interpretation of facts is dependent on assumptions (a picture, model, or organizing principle) brought to the facts in order to make sense of them.

8. Future discoveries could prove them false or of limited applicability.

9. Alternate theories might also explain the experimental results and observations.

10. Natural causes only

11. The assumption of uninterrupted uniformity—that what is observed in the present is what has occurred in the past.

12. A stipulation (to exclude consideration of supernatural causes) is taken as a conclusion.

13. (A discussion question)

14. A stipulation (to exclude consideration of purpose) is taken as a conclusion.

15. (A discussion question)

Chapter 7

1. c
2. b
3. f
4. e
5. a, c, d, e, f
6. e, f
7. f
8. e
9. e
10. It raised timeless questions in which everyone has a stake or at least an opinion: questions over conflicting perceptions of the meaning of life, between individual and majority rights, between academic freedom and parental rights, and between science and religion. The trial also set two highly public personalities at variance. A factor left unmentioned is that the trial set urban progressivism against small-town conservatism.
11. a, b, c, d
12. f
13. a, d, e
14. d, e
15. b, c
16. b, c, e
17. f
18. b
19. e
20. b, e
21. d

22. The claim seems credible because of an unstated philosophical assumption or predetermination that nature is all that exists.

23. (A discussion question)

24. c

25. a, b, d

26. Between 1809 and 1859, it was a philosophical idea with scientific pretensions, but one for which scientific evidence was lacking. Evolutionary change is always portrayed as being in the direction of increasing order and complexity, never decreasing. Evolutionary thought permeates disciplines outside of science such as philosophy and world history, as well as psychology and sociology (which, in order to be regarded as sciences, must assume causation and exclude intention). It permeates these disciplines to the degree that major disruptions to textbooks in such disciplines would result should evolutionary theory be thrown out. Evolutionary theory depends on extrapolations of the empirical data on species variability. The assumptions are that only natural causes were present in the unobserved past and that increasing specialization and ordered complexity occur naturally—assumptions that either cannot be or have not been empirically verified.

27. (A discussion question)

28. (A discussion question)

Chapter 8

1. CE, RI, CE, RI, CE, RI, RI, RI, RI, CE, RI, RI, RI, RI, RI, CE, CE, RI, CE, CE, CE

2. f, i

3. T, F, F, F, F, F, F, T, T, F, F, T

4. b, c, d, e

5. e

6. Any denial of objectivity in language is advanced as objective and violates the law of noncontradiction.

7. b, d

8. Yes. The only means to validate rational thinking would be through rational thinking, which would be pointless.

9. c, d, e

10. f

11. It must arise supernaturally.

Chapter 9

1. f
2. a
3. Origin, meaning, morality, and destiny
4. It would seem that the answer to the origin question should have a bearing on the answers to the other questions, but because science and religion are regarded as separate realms of thought, there is no connection between the answers.
5. e
6. e
7. b
8. God (or the supernatural)
9. Memory and testimony
10. An external reality exists. There is lawful regularity in that reality. That lawful regularity is rationally comprehensible.
11. T, F, F, F, T
12. e
13. T, T, T, T, T, T
14. e
15. methodology / philosophy
16. d
17. (A discussion question)
18. They must be assumed and, as assumptions, they are often merely arbitrary. Philosophers have searched everywhere for a reliable presupposition except the one place where they might actually find one.

Chapter 10

1. They never extend the logic as far as it needs to go, and no one else does either.
2. Any manmade object, or naturally occurring object in a state that is not its natural state, is proof that design is operative in the world. Much of that design seems intelligent.
3. c
4. As either a cultural or social phenomenon, or as a survival mechanism in some collective sense arising as a result of natural selection. Hidden objective values are likely communal harmony or common good in the case

of a cultural or social explanation, and survival and progress in the case of natural selection.

5. There is no design or value. Experiments must be designed and they are designed because the information expected from them is deemed valuable. They can only arise from some source outside of nature.

6. No insight or resolution is possible. In one case, the conditioned response is imposed by the social environment; in the other, by natural selection. But mere pressure to conform or impetus to survive does not establish rightness or truth. Reasoned arguments and value assessments are necessary.

7. The first is a universal statement; the second is a proposition advanced about reality.

8. No. Knowledge cannot be in a causal relation to that which is known without being, at the same time, subject to that which is known (as an effect is subject to its cause). Objective knowledge is characterized by propositional statements of affirmation or denial that owe their existence to logical grounds, not cause-and-effect processes.

9. Objective truth claims, the claim of naturalism included, do not arise as effects of natural causes.

10. c

11. That evil is a real problem, not an illusion. The existence of evil is objective, and objectivity is supernatural.

12. d

13. b (To infer from experience is to have already assumed rational order)

14. There is no meaning.

15. Purposes (or motives) are ranked according to their honorableness, morality is linked to individual and collective conscience, and hope that death is not the end persists across generations and cultures.

16. Is it relevant to common everyday experience of rationality, meaning, and goodness?

17. Attribute it to causation of some sort. Once a belief is established on grounds of causation, rational grounds become superfluous.

18. Freedoms and rights are values that become subjective when there is no authority to turn to in order to objectively establish them. Note that the Declaration of Independence, and, by association with it, the Constitution, places that authority in the Creator.

Chapter 11

1. Beliefs about that which is beyond nature can be a very divisive and contentious subject. The Bible stands as the sole exception—an instance of agreement and unity between the writings of two major world religions.

2. Their presence in the same book, despite the apparent divergence of the two religions

3. The Old Testament was not a Jewish invention.

4. Without knowing much about the specifics of either belief system, the falseness of the assumption is at least apparent from Jewish hostility toward the establishment of Christianity.

5. From Acts 23:6, 24:21, and 26:6–7: "With respect to the hope of the resurrection of the dead, I am on trial before you this day … And now I stand here on trial for hope in the promise made by God to our fathers, to which our twelve tribes hope to attain … And for this hope, I am accused by the Jews, O king!" (RSV) For Christianity, the resurrection turned out to be a chief point of departure from Judaism.

6. Jesus was a mythical character and not a real person. It is absurd to think that such a myth or legend could develop within the lifetime and geographic vicinity of those who would have known better. There are also nonbiblical references to his existence and from hostile witnesses (e.g., Tacitus). He did not die on the cross, but revived later on. The record is clear that Roman executioners were too effective for that to happen. There was a mix-up in tombs and the women went to the wrong tomb. All that the Jewish authorities would have had to do to settle the matter would have been to identify the correct tomb. The disciples stole the body and then lied about his resurrection. The disciples were demoralized and afraid of what might happen to them. It was only the certainty of the resurrection (including their own resurrection) that enabled them to proclaim it as consistently as they did in the face of severe opposition.

7. As a claim, it was highly vulnerable to being disproved; one would need only to have produced the corpse. The resurrection squarely confronts (and triumphs over) the central problem of life on this earth: death.

8. No other religion or belief system has confronted the problem of death with a demonstration of power over it.

9. b, d, f

10. (A discussion question)

11. (A discussion question); the author believes it is a strategic error, if only because of the ambivalence regarding the intent or purpose of any particular natural object or phenomenon, along with the consistent failure of such

arguments to settle the question. Notwithstanding the argument's historical merits, widespread belief in evolutionary science has certainly generated the perception that the argument for design from nature is vacuous. In addition, there are multiple claims to insight into the supernatural, which an argument from nature cannot settle. A three-faceted argument is advanced instead: (1) the consistency of a creator with rationality, purpose, and goodness; (2) the relevance of a creator to rationality, purpose, and goodness; and (3) the empirical evidence in the testimony of the Old and New Testament authors that the Author of Life would, and did, appear. However, it remains to be seen whether these arguments would fare any better against predispositions to believe otherwise.

Chapter 12

1. b, d, f
2. a, c, e
3. sight / reason
4. Reference to the Old Testament is frequently made to sustain arguments and points.
5. They testify how God has fulfilled the promises of the Old Testament.
6. reason / It sustains the presence of an unseen reality already verified as true, and supplies the motivation to overcome doubts and fears against acting in accordance with that unseen reality.
7. Irrational or immoral behavior may result. Yes, they can be. At least in cultures that have foundations in the Bible, they are ultimately deduced from the intrinsic value of individuals, a consequence of their bearing the likeness of the Creator.
8. No action taken on that which is known (by reason) beyond the realm of sight. Irrational desires, doubts, and fears.
9. integrity of character / propositional truth
10. No / When it is a venture to trust in someone / Yes
11. Lack of scientific evidence / Fear or contempt of self-appraisal
12. (A discussion question); while miraculous signs confirmed presentations of the gospel as Christianity was first becoming known, they would probably not be able to sustain a commitment to it. Continuing signs or appearances would probably be taken for granted after a time, or their impression would quickly be forgotten. Only reason and faith together are likely to sustain a consistent commitment.

SPECIAL SECTION:
Challenges Of The Skeptic

The present generation might be termed the "True for You but Not for Me" generation. In postmodern colleges and universities, the possibility of knowing any truth at all is questioned in the service of self-justification. It is always easier to raise these objections than to answer them. The following are some typical challenges by skeptics, along with a possible reply or replies to each. Note that it can be advantageous to expose assumptions by turning them back as questions. In this way, at least the initial potency of the challenge can be neutralized.

CHALLENGE 1: Ideas about God are just cultural devices to force compliance to arbitrary standards of behavior.

ANSWER 1-1: What about elevating tolerance as a god to force acceptance of deviant behavior?

ANSWER 1-2: Isn't elevating tolerance just a cultural device to unload feelings of conscience?

Elevating tolerance above other virtues is just as much a cultural device, and taken to its logical conclusion, such a move produces anarchy as the rule in human relations. Take away transcendence and everyone becomes his own authority on behavior. Empirically, that is not the way people behave; they behave as though there is an authority that transcends them. Tolerance is a virtue, but not the only virtue, and certainly not the highest virtue. It can and does come into conflict with other virtues. Parents discipline their children to conform them to certain expectations that they would not otherwise attain because they love them, not because they tolerate them.

The most radical skeptics think that religious superstition regarding wrongdoing is moral imperialism, a control and domination strategy. Advocacy of tolerance above all other virtues is the same sort of imperialism. Thus, there is a lurking contradiction: these skeptics are intolerant of anyone who suggests there are things that ought not to be tolerated. "Imagine the audacity of those who would dare to impose their beliefs on everyone else!" they say. But that is precisely what they do in requiring toleration of what they themselves say ought to be tolerated.

The word of God—the Bible—is the best proof of the existence of God and moral authority. However, it takes time to help people understand why that is so. Unless people are open to that possibility, a more expedient reply is to identify their assumption, stand that assumption on its head, and toss it back. In this case, that means showing them that the assumption that everyone should tolerate is also a cultural device.

CHALLENGE 2: The Genesis creation story is nothing more than a myth or fable.

ANSWER 2: What about the evolutionary "fish" story (fish becoming amphibians becoming reptiles becoming mammals becoming man)?

The credibility of Genesis 1 is grounded on the credibility of the Bible as a whole; that is, on its overall consistency and the verifiability of its other content. The credibility of the fish-to-philosopher story depends on assumptions that cannot be or have not been scientifically validated: natural causes only in the unobserved past and naturally occurring specialization and ordered complexity.

CHALLENGE 3: "True for you, but not for me."

ANSWER 3-1: What you are saying might be true for you, but how do I know it is true for me?

ANSWER 3-2: What you are saying might be true for you, but not true for me. So what is true for me is true for you also.

ANSWER 3-3: If truth does not exist for you, does that mean you're a liar?

Relativism is attractive, but the relativist has a rather awkward and uncomfortable alliance with truth. He wants relativism to be true in the most absolute sense, but any other proposition on matters of right and wrong is declared to be subjective and personal. A relativist wants flexibility to be relative on everything except the truth of relativism.

I once encountered an instance of a relativist objecting to someone else arguing a position on a controversial issue. That seemed rather self-defeating. It becomes impossible to object to someone arguing in favor of a view on a matter without, at the same time, asserting that one's own view, even if only the objection, is absolutely the right view.

CHALLENGE 4: What right do you have to judge other people's behavior or impose your morality on them?

ANSWER 4: What right do you have to impose your morality by saying it is *wrong* to judge someone's behavior?

This "question" of the skeptic is just as much a "thou shalt not" as any other. If a moral "opinion" ought to be kept a private matter, then theirs should also. This, of course, is impossible.

CHALLENGE 5: All religions are just different paths to the same end.

ANSWER 5: The only place a single end among multiple religions has ever shown up is in the Old and New testaments.

CHALLENGE 6: How can a good god exist with so much evil in the world?

ANSWER 6-1: How can evil exist without good existing also?

ANSWER 6-2: How can freedom exist without the possibility of choosing evil?

ANSWER 6-3: Are you expecting me to agree with you about evil in the world?

(Return a question as a means to introduce the idea of objectivity in such matters.)

CHALLENGE 7: As ridiculous as the propositions of Christianity are, how can the Christian worldview offer satisfying answers regarding our existence for which better ones cannot be found elsewhere?

ANSWER 7: The mystery of existence is tied up in the meaning of existence. Every worldview that attempts to establish meaning from within the world makes the world an end in itself. These views all end in futility because, sooner or later, death nullifies all sense of meaning derived within the world. If any ultimate meaning is to be found, the source of that meaning must be found beyond the world, in a view that suggests the world is a means to some larger purpose. Christianity is the only worldview that sustains a transcendent meaning to the world because it is the only worldview that confronts death and wins—and proves it.

CHALLENGE 8: Evolution is science; creation is religion.

ANSWER 8-1: Evolutionary science depends on assumptions that cannot be validated by scientific methods: natural causes only in the unobserved past and naturally occurring disorder to order. It would seem that evolution is not as scientific as one might think.

Historically, one of two approaches is to expose the lack of scientific evidence in support of evolution and so remove evolution from the science category. The other approach brings creation into the category of science by presenting and explaining the physical evidence in that light. History shows that neither of these approaches has worked well. The arguments in the main book do not attempt to categorize or de-categorize either as science. The definition of science is left to scientists. Instead, the arguments show that evolution is an ideology and creation is fact.

In lieu of the above reply, which identifies the unscientific assumptions of evolutionism, a less direct counter that cannot be answered without acknowledging that there is something beyond natural causes might be effective such as:

ANSWER 8-2: How can evolution as a proposition be true and creation as a proposition be false if both are inevitable products of natural causes?

A third possible counter to this challenge is to bring to awareness the fact that science depends on a universal rational order that cannot be known by scientific methods and that can only be accepted and maintained by faith. Science and faith (and thus religion) cannot be partitioned.

ANSWER 8-3: How do scientists know there is a universal rational ordering in nature when the reach of scientific observation and experimentation is limited?

CHALLENGE 9: Can God make a rock so big that he cannot move it?

ANSWER 9: Can you make a house so sturdy that you cannot tear it down? (There is nothing quite like posing an irrational and self-contradictory intention and demanding a rational explanation for it. Turn the challenge back on the questioner.)

CHALLENGE 10: Miracles violate the laws of nature, so they cannot happen.

ANSWER 10: Manmade objects do not exactly follow laws of nature either, so does that mean they cannot happen as well?

There are plenty of items that, while never attributed to laws of nature, are not violations of laws either. All manmade objects qualify because manmade objects would not exist if there were no interventions to disrupt and redirect the course of nature. In the same fashion, divine interventions may disrupt and redirect the course of nature, but they do not necessarily violate any laws. Miracles simply go beyond the limit of what disruptions and redirections human beings are able to impose on nature.

Clarence Darrow's questioning of William Jennings Bryan during the Scopes trial was intended to make individual miracles reported in scripture appear ridiculous, and it worked. Darrow tried to provoke Bryan to come up with naturalistic explanations, but miracles cannot be explained by the cause-and-effect operations of nature. Manmade objects cannot be explained by the cause-and-effect operations of nature either. At some point, the skeptic has to grant an unnatural (i.e., miraculous) origin for the creature that produced the objects.

CHALLENGE 11: Those who claim to know without a doubt that a creator exists are terminally arrogant.

ANSWER 11: What about those who claim to know what other people cannot know?

CHALLENGE 12: Invoking the supernatural as an explanation of something that science is still working on is simply explaining it away.

ANSWER 12: How is the promise that science may one day figure it all out any better?